Manufacturing Process Design and Costing

Simmy Grewal

Manufacturing Process Design and Costing

An Integrated Approach

 Springer

Dr. Simmy Grewal
Simsoft Knowledge Systems Pty Ltd
Sydney, Australia
e-mail: sgrewal@simsoftks.com
www.simsoftks.com

ISBN 978-1-4471-5719-9   ISBN 978-0-85729-091-5 (eBook)

DOI 10.1007/978-0-85729-091-5

Springer London Dordrecht Heidelberg New York

British Library Cataloguing in Publication Data
A catalogue record for this book is available from the British Library

Cover design: eStudio Calamar S.L.

Printed on acid-free paper

Springer is part of Springer Science+Business Media (www.springer.com)

This work is dedicated to the students and practitioners of Manufacturing Engineering

Preface

Process design is at the heart of all activity, it enables the manufacture of complex products like aircraft to something as simple as preparing a meal at home. This has led to the development of methodologies that aim to capture and formalise process knowledge. For the area of manufacturing process design variant and generative types of methodologies have emerged but the ability to apply process knowledge in an interactive and intuitive manner remains elusive. Those familiar with manufacturing therefore propose a heuristic approach, which is defined as a method of obtaining solutions via exploring possibilities rather than applying sets of rules or algorithms to solve a specific problem. A research project was undertaken to scope this possibility and what emerged from the effort was software that proved to be of value to industry and for teaching in universities, it is outlined here in detail.

Chapter 1 provides a perspective of product and process design illustrating the main steps involved. Chapter 2 examines the practice and procedures available and highlights the need for a heuristic approach. Chapter 3 details a new concept for process design based on the parsing of process narrative to separate the key variables involved and to determine their inter-relationships. The concept was engineered into a data schema that unifies part manufacture and assembly planning, hence creating new possibilities for costing and process knowledge management. Chapter 4 demonstrates the methodology through a manufacturing example and shows how it can be utilised for integrated manufacturing process design and costing. Chapters 5 and 6 provide tutorials for students and Chap. 7 focuses on industrial case studies. For further details visit the website http://www.simsoftks.com.

Sydney   Simmy Grewal

Contents

Chapter 1
Product and Process Design

1.1 Introduction

Product and process design is at the heart of all business activity and integration
with costing enables better decisions to be made and this helps to reduce business
risks. Such integration is still elusive and it is due to lack of a unifying methodology. The development of such a methodology requires effective integration of
the key variables and a user-friendly interface. Advances in computer systems and
software now allow us to attempt this and a research project was therefore
undertaken to study this possibility. What emerged from the effort was software
that proved to be of value to industry and for teaching of manufacturing engineering in universities. The concepts underlying this software are outlined here
and we commence by looking at the nature of product and process design.

1.2 Product and Process Design

In order to determine the manufacturing cost of a product and quickly assess a
'what if' scenario in terms of a change at the product, part or assembly level, an
integrated perspective of product and process design is required. Such a perspective is provided by Fig. 1.1.

1.2.1 Product Planning

Product planning remains complex due to the interactions involved between the
customer's requirements and the product's functional attributes. Concepts have
been developed to assist this process and references [1–3] highlight some of them.
Product life cycle requires consideration due to shortening of product life in the
marketplace and emissions requirements. It can be viewed from different

S. Grewal, *Manufacturing Process Design and Costing*,
DOI: 10.1007/978-0-85729-091-5_1, © Springer-Verlag London Limited 2011

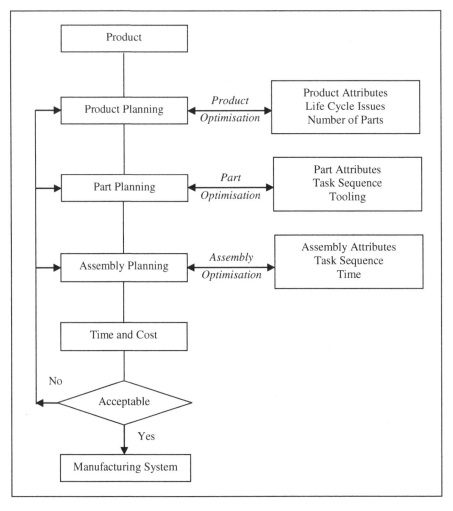

Fig. 1.1 An integrated perspective of product and process design

perspectives. From the manufacturer's perspective it is the time from the conception of product to its final withdrawal from the marketplace. From the marketing perspective it is the growth, maturity and decline of sales. From the customer's perspective it is the purchase of the product to its final disposal. In reality, it is from the conception of the product to its final disposal regardless of other stages.

Product design is broken down into its constituent parts in order to make the manufacture possible, these parts then require manufacturing and assembly. This creates a number of problems and the foremost among them is the number of parts. If the number of parts can be reduced by even a small amount then the benefits cascade down into all the activities that follow. This leads to significant cost savings in manufacture and increases the product's reliability as there is less to go wrong. To assist parts rationalisation various concepts have been developed and they form the basis of design for manufacture and assembly guidelines [3]. These concepts have been applied in industry to streamline design but the increasing functionality and the sophistication of products is taking us toward more and more parts and this requires new approaches for parts rationalisation, such as those based on costing [1]. The methodology outlined in this monograph focusses on this.

1.2.2 Part Planning

Part planning has been the domain of those well versed in manufacturing engineering. It is a skill-based activity and the specialists often arose from the factory floor and brought with them considerable heuristic knowledge of processes and equipment involved, such as jigs, fixtures and machine tools. The author went through this process and gained a great deal of knowledge about part manufacturing activity. The details in part planning are of technical nature and they commence with the material and the volume involved. These variables influence the manufacturing process and tooling and generate the macro-aspects of part planning. In these macro-aspects are inherent the micro-tasks that help to create the part form. These are the shaping processes, such as milling and turning, and they constitute the micro-aspects of process planning. The macro-layer can be generated as a sequence of tasks and these tasks then analysed for their micro-requirements as shown in Fig. 1.2.

The variables influencing the manufacturing cost are materials, equipment and the tasks involved that create the part form. Cost is locked in as soon as the part form is finalised and every aspect of it after that influences the cost of manufacture, particularly the surfaces to be generated and the tolerances to be met. Once the manufacture starts the cost starts to build up from the amount of material involved including the scrap amount. To process the material special equipment is often required and this brings in their cost of utilisation as shown in Fig. 1.3. Manufacturing expertise is applied through the analysis of tasks and this involves the setting up of the process. This generates the total time required for the process including allowance for inefficiency, which is always present. The costing of time is another matter and it varies considerably depending on the location of manufacture, especially the country of manufacture. In the recent times this is evident from the relocation of manufacture to China and India. The time involved in the manufacture of a product is same everywhere but the cost is not, hence the ongoing effort to reduce it by shifting location around the world and this trend will continue in the foreseeable future.

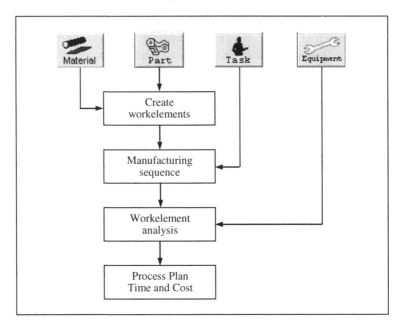

Fig. 1.2 Part planning

1.2.3 Assembly Planning

Assembly planning is no different from that of part planning, the task analysis is once again the key requirement. Assembly process has a sequence of macro-tasks and their analysis leads us to the equipment and time requirements as illustrated in Fig. 1.4. The macro-sequence of tasks helps us to establish the overall assembly process and this starts with the first component involved and finishes with the final task. This task-based commonality provides the underlying unity to part planning and assembly planning, this can be utilised to capture the overall manufacturing information content of a product. This is a holistic approach to product design because it is based on total cost rather than number of parts. The use of standard parts instead of a single unified part can sometimes reduce the cost of manufacture and this requires an overall perspective of manufacturing. In assembly the cost of parts is brought in by the Bill of Materials (BOM), after that the cost model does not differ much from that for part manufacture as illustrated in Fig. 1.5. The final cost in this case reflects the total cost of manufacturing the product, including assembly. What follows after part planning and assembly planning is the design of a physical system to produce the product in volume. For this the process designs for part manufacture and assembly become the inputs.

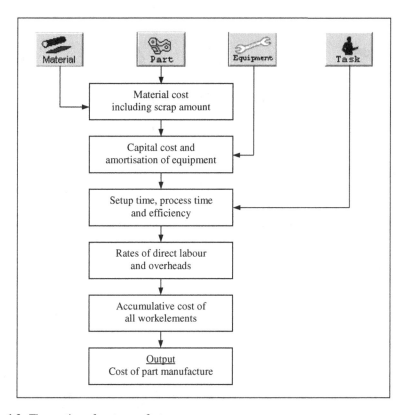

Fig. 1.3 The costing of part manufacture

1.3 Manufacturing System Design

Whether the product is manufactured in volume or as a one off the task analyses of part manufacture and assembly require a physical system to produce it. For this the task models become the input for the manufacturing system design as illustrated in Fig. 1.6.

1.3.1 Workstations

Volume manufacture requires concurrency of tasks and in physical systems this is provided by the workstations. The number of workstations is determined by the volume to be produced, shift time and the total time of overall tasks. This leads to the cycle time per station.

Often it is impossible to aggregate the task times involved to be exactly the same as the cycle time, some stations therefore end up being less than fully

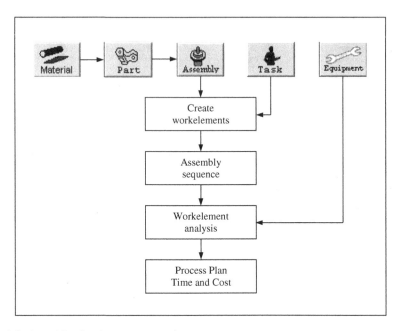

Fig. 1.4 Assembly planning

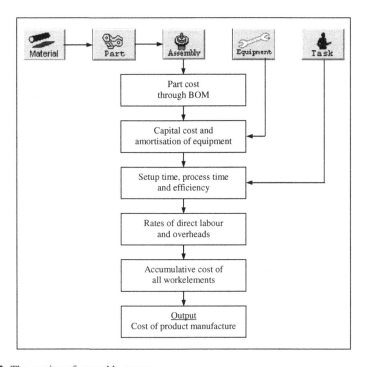

Fig. 1.5 The costing of assembly process

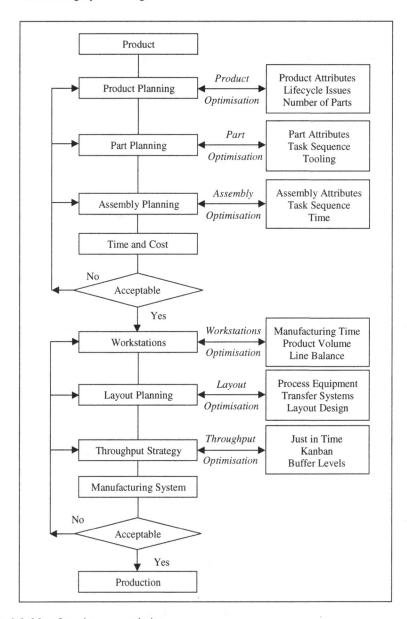

Fig. 1.6 Manufacturing system design

occupied and this leads to inefficiencies. The cycle time reflects the throughput rate or how many will be made per hour or per day, as for example in a bakery or in a car plant. If the production requirement is very high then the cycle time can be significantly less than the smallest task time. This leads to problems which are overcome by parallel workstations performing the same task. This helps to reduce the task time by producing more in the same time through concurrent activity. This application of parallel tasking at micro-levels helps to solve manufacturability problems to meet high production rates. If the task time is in seconds then a dedicated automation is often the only answer. In more complex products manufacture in high volumes the financial and the human resource issues become much more critical and generally involve high levels of business risk.

1.3.2 Layout Planning

Workstations require layout arrangements in order to meet the demands of equipment and transfer systems. In high volume manufacture layout becomes a significant part of the overall system design and transfer lines are examples of this. They involve machining cells and automated movement of parts and subassemblies, effectively the overall task models of parts manufacture and assembly are mechanised. One important point to note here is that in such settings the dynamic aspects of process design also become very significant. The static picture of macro-tasks and their micro-analysis does not bring to surface the dynamics involved or the mass and motion effects of parts and assemblies. A transfer line in full motion is a highly dynamic system, it is a process design in motion. This brings into play many other aspects of mechanical design and control systems which are beyond the scope of this monograph. One important advance in recent times has been the robotics technology. It has opened up new possibilities through fixed platforms and autonomous systems. Although it is an advancement of numerical control systems, the dexterous capabilities of robotics allow automated transfers and this enables the layout to be considered in a new light. There was a weakness in the processing systems for parts manufacture and assembly centering on the handling systems and this has been addressed by the robotic technology.

1.3.3 Throughput Strategy

The need for continuous flow of production resulted in just-in-time type of manufacturing, which in turn lead to large supply chain systems involving several countries. Such large systems are sensitive to unforeseeable circumstances that can delay the delivery of parts and assemblies. To overcome this buffer levels were created and there was a time when such buffer levels used to be very significant, until it was realised that this involved tying up large capital that could be more effectively used. This led to lean manufacturing which is an extension of just-in-time manufacturing in order to reduce the buffer levels to minimum or to eliminate them completely. The manufacturing company that took the lead to introduce this was Toyota of Japan, hence the just-in-time type of manufacturing is often called the Toyota System. Now even non-volume manufacturers, such as aircraft manufacturers, are applying such concepts to improve the productivity of their working capital.

In summary, process design translates product design into manufacturing requirement and this leads to time and cost of manufacture. It is about establishing the macro-tasks involved and to determine their micro-requirements. There is an underlying unity to part planning and assembly planning and this centres on the need for task analysis in both cases. This unity can be leveraged for integrated manufacturing process design and costing. In the following chapters we look at this in detail.

References

1. Otto, K., & Wood, K. (2001). *Product design—techniques in reverse engineering and new product development*. NJ: Prentice Hall.
2. Prabhakar Murthy, D. N., & Blischke, W. R. (2005). *Warranty management and product manufacture*. Springer, UK.
3. Boothroyd, G., Dewhurst, P., & Knight, W. (1994). *Product design for manufacture and assembly*. Marcel Dekker, New York.

Chapter 2
Practice and Procedures

2.1 Introduction

Author recalls a comment made by his manager in the early 1970s to the effect that "…I wish somebody could tell me exactly how much this part will cost to manufacture…" The machine tool company in the UK where the author was trained and subsequently employed as a Methods Engineer had been in the business of making large vertical turning centres since 1887 [1]. The question arose because we were trying to establish the best possible route to manufacture a certain part and this required access to many strands of information which were in different departments. The practice at the time was that someone did the methods engineering and someone else did the time estimations, costing was done at global levels, hence there was no way of getting at the individual part manufacturing cost, it was too inaccessible.

In more recent times a Blue Book Series 2003 from CASA/SME on 'Cost Engineering: The Practice and the Future' makes an interesting comment on the state of art at the beginning of twenty-first century [2] and makes a following remark: "Another area of future growth and research in cost engineering is to capture and reuse human expertise or knowledge used during the development of a cost estimate. This micro-knowledge management will help to analyse an old estimate better before it is reused. The current commercial software needs to go a long way to develop this capability in an intelligent manner so that the additional workload on the cost estimators is reduced." The basic task of methods engineering is to design the most economical route for part manufacture and assembly. The way the variables are handled and integrated has changed, more through an evolutionary process than a revolutionary one, as a result more transparency is now possible. The overall capability still remains short of the accurate and instant answers required by the professionals to make rapid decisions. The strong need and the immediacy to answer a cost-related question will never go way, after all business is about making profit otherwise it will not survive, we look at the practice and procedures available to assist this process.

S. Grewal, *Manufacturing Process Design and Costing,*
DOI: 10.1007/978-0-85729-091-5_2, © Springer-Verlag London Limited 2011

2.2 Practice

What is actually involved in part manufacture and assembly planning? What are the variables and how do they interact? What is the nature of costing and is true costing really possible? What roles computers can play and what are the limitations of people involved? We attempt to answer these questions in order to better understand the practice involved.

A methods engineer develops through training and practice an intuitive feel for the processes involved in part manufacture and assembly. When he (he is used in a general sense throughout this monograph to refer to a person) sees a paperclip he perceives all the processes involved, commencing with the feeding of the wire into the manufacturing system, cutting of it to length without influencing the material properties, the bending of it into the required shape without spring-back effect, and the packaging of it into container for delivery, just to mention a few of the macro-tasks involved. On the other hand, when an ordinary person sees the paperclip he sees the form and the function only, he cannot discern the processes involved in its making, just like seeing cooked food but not knowing the process involved in its making. The manufacture of complex parts, such as gears, is such a difficult task that it is a speciality in its own right. This complexity centres on the material properties, the shaping processes, quality and handling attributes, and it is made more complex by volume requirements. This is just for one part, given that there are often many parts in products, the activity of process design soon becomes a demanding exercise and requires procedures to assist it.

The above examples illustrate the fundamental issue in process design, it is that form creates relationships and these relationships have to be management. The correct mapping of these relationships into the data is at the core of process design, and manufacturing knowledge plays an important part in this. The relationship can be one-to-one, one-to-many or many-to-many. A part can be planned by one person who has all the required knowledge, example of one-to-one, a more complex part may require the input of several people, an example of one-to-many, or many different parts may have to be planned by many different people, an example of many-to-many. This creates complexity that soon becomes unmanageable and given that there are also many other relationships that require detailed considerations and things often change unpredictably we end up in a scenario where process design and the management of it becomes a demanding exercise. While the author was carrying out a cost-modelling project in China [3] the vice president of the company involved remarked that even the most predictable cost that of materials has now become seriously unpredictable, because as a commodity it has become part of the futures market, a cost modelling done at certain stage therefore becomes invalid even few hours later. The processing equipment have a value at certain point in time based on their depreciation rates, if this is not correctly accounted for the results will be wrong. Similarly, if a more experienced resource is replaced by a less experienced one then the costing of part manufacture will be wrong, this leads to some of the difficulties involved in the costing of manufacture. It is valid at a certain point in time only, just like a balance sheet, and this point of time is often different from that when the part is actually manufactured and assembled. It may appear therefore it is an intractable problem and at best we can live with the estimates only. To overcome this requires live or real time relationship among the variables involved so that they could be manipulated at will and at any time to provide a snapshot of true cost. This was not possible until the recent time when low cost computers came along and software capabilities emerged to help minimise the influence of time.

2.2.1 Methods Engineering

Product design requires planning for manufacture; otherwise it remains an idea only. This planning is the genesis of process design and it has been there from the beginning when industrial manufacture commenced in earnest. The metals-based mass manufacture commenced at the onset of twentieth century and led to the development of methods engineering discipline. Prior to that manufacture was a craft-based activity and an individual often carried out all the processes involved to produce the product. He manufactured all the parts and then assembled them and had all the procedural knowledge required, which was often gained through long apprenticeship. There was no need for him to communicate his knowledge, the procedure existed in his mind only and he continuously improved it through practice. The mass-manufacturing activity changed all that and created the need for communication of process design to others, this was the beginning of the practice of process design. A 100 years on we are still grappling with the complexities involved, particularly in the formalisation of process knowledge and its communication.

In the beginning communication of process design knowledge was via sketches and they depicted the actions involved, this method was easy to understand and is still in use today, you only need to look at self-assembly instructions of a product. This approach worked well for assembly because to describe the process involved would be unbearably lengthy, this is not the case in part manufacture. A drawing or a sketch of part is the starting point, what is required after that is the description of steps and the tooling involved. This led to the evolution of methods sheet [4] which records the sequence of events as a narrative. The development of this narrative and the interpretation of it require great deal of manufacturing knowledge and this led to the discipline of Methods Engineering. This discipline has served well for over a century to assist the manufacture planning. The methods sheet captures the process knowledge, therefore much effort went into its development. We look at the procedures that evolved from it.

2.3 Procedures

2.3.1 Constructive Method

The nature of form is complex, it can be made up of straight lines or curved lines, or it can be two-dimensional like a flat sheet, or it can be three-dimensional like a solid casting. These forms require accurate interpretations for manufacturing and this led to the manual approach for process planning. The methods engineer who is familiar with the resource available to carry out the task first examines the drawing of the part in detail and then writes down the steps involved, including the details of machinery and the tooling involved. This leads to a routing sheet for part manufacture and the procedure is called the Traditional Method [4]. With the evolution of computers it became possible to store and retrieve the routing sheet and the combination of manual interpretation of drawing and the utilisation of computers to manage the data led to the Constructive Method [4]. This method became the forerunner of more advanced techniques.

2.3.2 Variant Method

It was inevitable that computers will be utilised to exploit the similarity of parts and this led to the Variant Method [4], it became the cornerstone of what is we now call Computer Aided Process Planning (CAPP). Variant Method exploits the similarity of parts as well as the similarity of processes. This similarity can be on the basis of individual process or it can be on the basis of group of processes. The main disadvantage of this procedure is that the quality of process design still depends to a considerable extent on the knowledge of the methods engineer. There are no inference mechanisms in the procedure to fine-tune the process design. This is often required because there are tolerances and various levels of stresses involved in the manufacture of parts and this requires experience to create a feasible process plan.

2.3.3 Generative Method

Computer-aided automation of process design requires a whole new approach to part manufacture and assembly planning. This centres on the decision logics, formulae, algorithms and data extractions. This approach is labelled as the Generative Method [4]. Examples of such procedures are Decision Trees, Decision Tables, Axiomatic and Expert systems. Regardless of their approach they remain complex in nature and this led to the specific applications, such as those in sheet metal fabrications and electronics manufacture, where parameters are more confined. The development of computer-aided drawing and solid modelling has now opened up new possibilities for features extraction, but the overall automation of process design remains difficult; this has more to do with the ability to foresee difficulties in part manufacture and assembly than just meeting the requirements of micro-processes.

2.4 Costing

Costing requires process plans as an input and the procedures described above provide for this. Costing activity used to be called cost estimating, but is now generally termed cost engineering [2]. This encompasses other functions such as decision-making and budgeting by predicting the cost of activities, we look at the practice and procedures in this area.

2.4.1 Traditional Approach

Decades ago, products were simple in their design and the number of parts involved was low, this simplicity enabled the costing to be done by an individual and cost-estimating became a specialised area. The cost estimator gained considerable know-how of the costs involved and could give first-cut estimates of new products or family of them in a rapid manner. These first-cut estimates were not far off the mark, later when the detailed design emerged it was possible to validate the cost estimate by examining the nature of manufacture involved. These detailed estimates looked at the activity times and the associated costs and this led to what is called activity-based costing. It is a useful concept because it is based on quantitative analysis and is linked to the hard data on manufacture. The drawback of this concept is that the stage at which this is done is often far too downstream of the product conceptualisation stage, therefore limiting its potential for new product designs. The products of today are very sophisticated in terms of their number of parts and the technologies employed and it has become imperative to know the cost as soon as possible in order to reduce the business risk, this has led to other procedures for costing.

2.4.2 Parametric Approach

The parametric approach utilises the similarities of variables as for example in the building industry. The cost of building is often related to the square metres of surface involved and this means the amount of work and the cost of materials is known from previous experience. Such similarities are valid in most cases because in the business environment there is much emphasis to minimise the risk. This is achieved by a gradual improvement of design; in other words products from one stage to next involve only incremental change and this can be readily handled. It is generally assumed that the change is linear and the costing can be extrapolated from one product to another, as long as there is a similarity of design. This linearity can be expressed mathematically and utilised in cost engineering. This may appear simplistic but it has certain benefits, for example it is of much value at the product conceptualisation stage, this means an early indication of manufacturing cost is known and this reduces risk.

2.4.3 Integrated Approach

The integration of costing with process design is desirable but often it is not possible. The features based costing provides possibilities, through features the part is described as a sum of standard features such as faces, edges and holes that require manufacturing. The cost of manufacturing a feature is first established and then used across other parts. For this to be effective the manufacturer requires a large geometric database. This presents problems because the definition of a feature is not so easy to standardise across organisation even though it may be manufacturing a similar class of product. To overcome this research effort has turned to neural-networks which are part of artificial intelligence systems. The aim is that computers can learn the relationships involved among the attributes and the costs. In the end analysis we go back to the point made earlier that there are infinite forms to be produced, hence there are infinite process design possibilities, and a heuristic approach becomes particularly attractive in such settings. It is defined as a process that obtains solutions by exploring possibilities rather than by following sets of rules or algorithms. The activity of part manufacture and assembly planning is the interplay of materials, form features, equipment and the task attributes and planner achieves optimisation through intuition and experience and this makes the process akin to a heuristic approach [3].

References

1. http://www.webster-bennett.co.uk.
2. Roy, R., & Sackett, P. (2003). *Cost engineering: The practice and the future. Blue Book Series 2003*. USA: CASA/SME.
3. Grewal, S., & Choi, C. K. (2005). An integrated approach to manufacturing process design and costing. *Journal of Concurrent Engineering: Research and Applications*, 13(3), 199–207, SAGE Publications (ISBN 1063 292X).
4. Groover, M. P. (2002). *Fundamentals of modern manufacturing: Materials, processes, and systems* (2nd ed.). New York: Wiley.

Chapter 3
Schema Design

3.1 Introduction

A methodology or a procedure requires a schema; it is defined as the principle that enables the understanding to unify experience. The design of a schema is half the problem, the other half is the user interface to it. We look at these requirements by examining the nature of process design and how it is communicated. The word 'process' is used quite frequently in everyday communications to describe events that help to create something useful or achieve some outcome. In the business environment it defines the events that create value adding; in the manufacturing environment it defines the method that creates products; in the cooking environment it defines the recipe. This widespread use of word shows its all encompassing influence as a means of expressing activity and this leads us to the question whether it is possible to design process along scientific lines rather than evolve it through trial and error, which is often the case, witness the comment when something goes wrong, it is invariably the process that needs looking at.

Without the process food will not grow, human cannot be born, aircraft or a car cannot be manufactured, hence its all pervasive influence in our daily lives. Whereas the food or a child or a manufactured product is a tangible object, the process involved in it is not. For example, in a paper clip an ordinary person cannot discern the steel making process or the manufacturing process. This is because process requires ingredients of time and action and they get encapsulated in the final outcome and are therefore not discernable to ordinary people, whereas a specialist understands the variables involved and sees the object differently. Variables and their relationships provide the structure to process design and to understand this we need to look at how process information is communicated. Language of course plays a key part in this and it is used to paint word pictures, these pictures come together to form process knowledge.

S. Grewal, *Manufacturing Process Design and Costing*,
DOI: 10.1007/978-0-85729-091-5_3, © Springer-Verlag London Limited 2011

3.2 Process Knowledge

In manufacture the common denominator is task and process is expressed through task. The tasks are the external expressions of process and they reflect the process knowledge involved. At the highest level process knowledge is expressed as words and we use words to define actions that are invariably performed on some objects. Actions and objects come together to form the process narrative, which is at the core of process design. Let us look at an example:

Add Sugar; Pour Milk.

We are familiar with these expressions as being related to the making of tea or coffee, we need to, however, look at them from a different perspective. In the instructions we find that the actions are Add and Pour, and the objects are Sugar or Milk; the former are verbs the latter are nouns. This "grammatical parse" of the narrative to isolate the actions and the objects involved helps us to bring to the surface the structural essence of process description [1]. We can rearrange the narrative to see more clearly the associations amongst the actions and the objects involved and the sequence of events, as shown below:

Add   ⟶   Pour
Sugar   Milk

There are several things that can be noticed in this arrangement. Firstly, the attributes of the objects influence the attributes of actions; secondly, the 'Add' and the 'Pour' actions have motion and equipment requirements. The profile of motion is dictated by the attributes of sugar and within this profile is inherent the quality of the task and also the time involved to perform it. The description 'Add Sugar' is a macro task, whilst the action profile is a micro task and the sequence of events ensure the precedence. This provides the mechanism to structure and visualise process information as shown below:

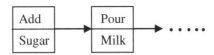

3.3 Process Information

Process information is the first level of visualisation of macro parameters; it lies midway between process knowledge and process data. The translation of process knowledge into process information is the main activity of process design; it translates the word picture into detailed action format. The issues after that centre more on data development and management. It is at the interface of process knowledge and process information that the heuristics offer the greatest potential, and at this interface we translate abstract knowledge into concrete information.

3.4 Process Data

Process design centre on the correct analysis of actions involved and this in turn is influenced by the attributes of object as illustrated in Fig. 3.1. We look at this data mapping process via a closer examination of the 'Add Sugar' task. The word Add as a verb is generic in nature and it is a high-level descriptor. The process description generally is at a high level, and is the only way of expressing process knowledge in an encapsulated form. In this is inherent one of the difficulties involved in formalising process knowledge. It is easy to express process knowledge at a high level, but it is not so easy to express it in a detailed manner. This is because the high-level description allows it to be used in different settings. As a word picture the Add action generates the image of something being added to something else, but used in the context of Add Sugar the image generated is of a specific nature, because we are in a position to see the activity involved from our past experience. The activity in this case centres on the sugar being lifted by spoon and added to the cup, it does not give us any feeling about the weight, the forces involved or any indication of the time involved. In real processes, attributes such as these play important role and influence the efficiency of the task. The process narrative by itself does not fully capture or express these aspects, but we can use it to develop parameters such as time and cost through detailed analysis.

In translating the process knowledge into the data, the amount of detail involved increases and the transparency decreases. To see this you only need to look at the assembly instruction of a consumer product or recipe details. Process narrative is everywhere, including the sequence of events, but the detailed mapping of object requirements into action requirements is nowhere, it is left to the experience of the assembler or cook to provide this detail. Another example illustrates this point: it is the tightening of nuts in the assembly of aircraft structures. The Tighten Nut instruction requires information on the serial number of wrench used, the torque settings of it, the date and the name of the person who last tested it. Given that there are thousands of nuts and bolts in commercial aircraft, the standardisation of process to its minute detail becomes necessary because of the level of documentation required in the manufacture of aircraft. The above examples illustrate the nature of relationship amongst the tasks and the objects.

Fig. 3.1 The mapping of object attributes into the task attributes

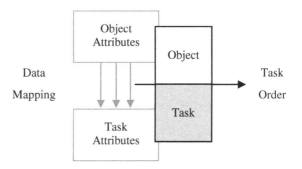

3.5 Data Relationships

The data relationships provide the structure for the schema design and we look at this in detail. As a general rule data relationship can be one to one, one to many, or many to many [1]. The complexity increases as we move away from one to one to many to many, and the latter is often the case in manufacturing as illustrated by the hierarchy of information shown in Figure 3.2.

Starting at the top, the assembly data object has 'one-to-many' type of relationship with the part data object, whilst part object has 'one-to-one' type of relationship with the assembly object. This determines the schema for the assembly object and bill of materials (BOM) is an example of it. The part object has 'one-to-one' type of relationship with the task sequence object and it is the same vice versa. In practical terms this means that there is always a unique sequence of tasks to manufacture the part dictated by the material and volume involved. The task sequence object has 'one-to-many' type of relationship with the task object and the task object has a 'one-to-one' type of relationship with the task sequence. This means that a task sequence often contains many different types of tasks. Finally, the task object has 'one-to-many' type of relationship with the material object and equipment object, whilst the material and the equipment objects have a 'one-to-one' type of relationship with the task object. In practical terms this means that a task often utilises many different types of materials and equipment to make the parts. The determination of these relationships together with the object attributes lead us to the schema design that can hold all the information required for the manufacturing process design. The data relationships model of Fig. 3.2 integrates the key variables involved and their interactions

Fig. 3.2 The data relationships in manufacturing

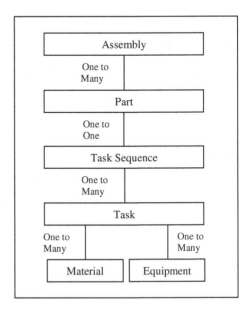

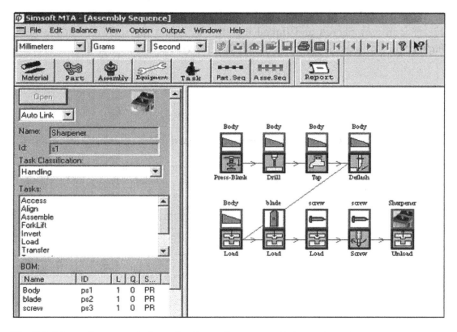

Fig. 3.3 The main menu window of software for process design and costing

determine the cost of manufacture. This data relationship model became the basis of software for the heuristic approach [2] and its main menu window is illustrated in Fig. 3.3.

3.6 Schema Design

The task analysis structure and the data relationships constitute the nature of the schema design. The task analysis structure has a visual aspect to it and the data relationships have a relational aspect to it; the former are part of the user interface design and the latter are of database design and enable the software engineering to be addressed. In this is also inherent the need to separate the application from the user file, so that many can use the software and manage their own files to safeguard their work. The relational model of Fig. 3.2 was engineered into a schema via linking the data tables together and Fig. 3.4 shows the 'Assembly' data object and

Name	CreateDate	IId	DimX	DimY	DimZ	Weight	PartQty	PartCnt
Sharpener (Aluminium)	2002/07/30, 1£	AS1	35	30	15	11	5	3
Sharpener (Plastic)	2003/05/25, 1⁴	PSA	35	30	15	8	5	3
			1	1	1	1	0	0

Fig. 3.4 The database model for the assembly object

the variables linked to it. In a similar manner the relationship tables for all the objects were engineered and linked together to create the software. In the next chapter we will look at the benefits this has provided.

References

1. Pressman, R. S. (2005). *Software engineering: a practitioner's approach* (6th ed.). UK: McGraw-Hill International Edition.
2. Grewal, S., & Choi, C. K. (2005). An integrated approach to manufacturing process design and costing. *Journal of Concurrent Engineering: Research and Applications,* 13(3), 199–207 (ISBN 1063 292X).

Chapter 4
Heuristic Approach

4.1 Introduction

A heuristic approach obtains solutions via exploration of possibilities rather than following a set of rules or algorithms to solve a problem. Solutions obtained by rules or algorithms are generally prescriptive by nature and are not flexible enough; a heuristic approach on the other hand provides an opportunity to explore various possibilities. In the manufacturing environment there is a need to manufacture many different types of forms and because the volumes also differ there are often many processes involved and they have to be carried out cost effectively, a heuristic approach therefore becomes particularly attractive in such settings. The process planner often knows the alternatives but the state of art is such that it is often not so easy to translate them into detailed cost. In the manufacturing environment issues arise almost instantaneously, for example a process machinery breaks down and an alternative routing is required to continue the production, this involves technology and cost assessments and they need to be done rapidly, a methodology that can help meet this need would be a valuable aid for the process planner.

The demonstration of an idea via a drawing or a sketch is relatively straightforward, this is not the case in process design because it is expressed as a narrative and this requires interpretations; it can have different meanings to different people. In order to overcome this the narrative is often accompanied by images and this helps to overcome any misunderstandings, it raises the question therefore whether it is possible to combine both means of communication in a single methodology to leverage their strengths. This was not possible until the recent times when the graphical interface to the computers developed and it is now a common feature to all types of operating systems. This development began in the late 1980s and led to the interest to leverage this opportunity for 'sketching' the process narrative in order to increase its transparency and accuracy. The author was appointed by the Australian Government research organisation CSIRO (www.csiro.org.au) in the late 1980s to study this possibility and to develop tools

S. Grewal, *Manufacturing Process Design and Costing*,
DOI: 10.1007/978-0-85729-091-5_4, © Springer-Verlag London Limited 2011

Fig. 4.1 A plastic bodied
pencil sharpener

for industry to enable better process designs and costing capabilities. What follows is the result of this effort and it is demonstrated through a simple manufacturing example that focusses on the part manufacture and assembly planning of a plastic pencil sharpener as shown in Fig. 4.1. This is a familiar product and the manufacturing issues are generic in nature and are equally applicable to more complex problems.

4.2 Part Planning Process

4.2.1 Part Attributes

Manufacturing deals with the transformation of materials into forms. This is a creative process and demands considerable know-how of engineering processes. The material and the part form have a considerable influence on the manufacturing process and this starts with the attributes of part, they enable the development of sequence of tasks that generate the final form.

The nature of material involved and the volume to be produced have the most effect on the method of manufacturing and this is followed by the tolerances to be met. The selection of materials in conjunction with the dimensions of part determines the cost of material as illustrated in Fig. 4.2. It is a software window showing the attributes of the sharpener body. The overall dimensions refer to the boxed dimensions of the body and these dimensions determine the handling attributes of the part in assembly. The plastic material is generally bought on the basis of weight and the weight of the body enables us to cost this including the scrap amount. There is always some scrap amount involved in the manufacture of parts and this has to be accounted for in terms of the material cost; some manufacturers cost this to levels as low as fractions of a percent.

A study of the other fields throws light on the nature of planning information. Some details are self-evident such as the name of the part and its identifier. The category number refers to the category the part belongs to and whether it is a machined part or a casting, etc. The icon selection from the library allows for the

Fig. 4.2 The data attributes
of plastic body

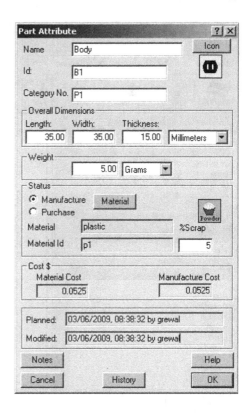

representation of the body shape on the planning screen. The status fields relate to
the part being either manufactured or purchased. In the latter case it is the purchase
price that is of relevance, as for example in the purchase of standard parts such as
screws, etc. The material cost and the manufacturing cost are same in Fig. 4.2
because we have not yet created the manufacturing task sequence for the body.

4.2.2   Manufacturing Task Sequence

In order to determine the manufacturing cost we need to design the manufacturing
task sequence, for this we must be able to translate the process narrative into the
structured information as outlined in Chap. 3. The software methodology allows
this to be done rapidly as shown in Fig. 4.3. By selecting the tasks involved from
the library on the left of Fig. 4.3 and by locating them in the main window as
shown leads to string of tasks and their precedence is maintained by the schema.
This sketching capability and the representation of task sequence via data glyphs is
the novel aspect of this methodology. The sequence is a macro-plan of events
capturing the primary tasks involved; it can be made as detailed as required or it
can be a broad-brush sketch, in this is inherent the flexibility of this methodology.

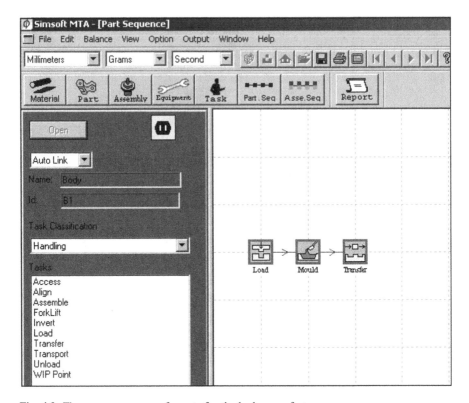

Fig. 4.3 The macro-sequence of events for the body manufacture

The requirement is often to 'paint' the general scenario rapidly and then refine it later. The sequence shown in Fig. 4.3 is for the plastic body; it can be copied and modified for metal-bodied sharpeners hence leveraging the database.

4.2.3 Task Analysis

Task has resource requirements and this is determined by the nature of the task and the volume involved. The nature of task is mainly dictated by the attributes of part, in other words the relationship the 'object' has with the 'action' involved. This relationship is data centred and it is possible to determine this through inference mechanisms. The process designer infers through experience, whereas software engineering on the other hand allows this to be done automatically. When task is associated internally with the object, the attributes of the task are interrogated in relation to the attributes of object. This interrogation allows for the selection of correct equipment based on the dimensional attributes of object. This mechanism can be extended to other resource requirements. By clicking the task, the resource

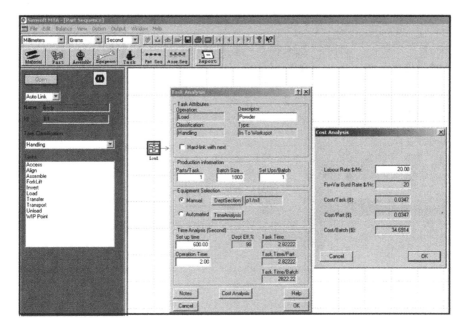

Fig. 4.4 The task analysis windows

requirement window opens up and allows the capture of task attributes, further attributes are captured through sub-windows as shown in Fig. 4.4. This completes the resource assessments, including cost.

After all the tasks have been analysed in this manner it becomes possible to integrate the information on time and cost and output the summary as shown in Fig. 4.5. This information incorporates all aspects of the part manufacture. When this part is used in the assembly planning it appears as a live relationship. Any changes in it are directly reflected in the assembly planning information and this enables the connectivity of all process information in an integrated fashion.

4.3 Assembly Planning Process

4.3.1 Assembly Attributes

Assembly is a 'part' constituted by other parts through Bill of Materials (BOM). Sometimes the parts constituting the final assembly are also assemblies themselves and they are called sub-assemblies. In large products the parts lead to sub-assemblies, which then become 'parts' in higher-level sub-assemblies. All the sub-assemblies eventually merge to create the final assembly or product. This is evident in all the sophisticated products and it is the only way to manufacture

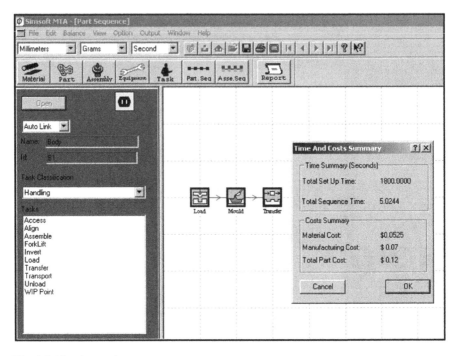

Fig. 4.5 The time and cost summary

Fig. 4.6 The assembly
attributes

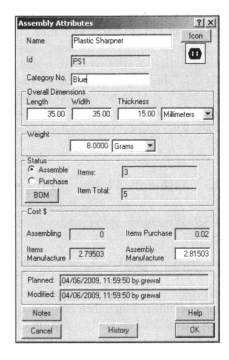

them. The sharpener as an assembled product shares attributes with the part attributes as shown in Fig. 4.6. The BOM window allows for the capture of all the parts that make up the assembly and this is reflected in the weight and item total fields of Fig. 4.6. The assembly cost appears after the assembly task sequence has been designed and analysed.

4.3.2 Task Sequence

Again we are able to simulate the assembly process narrative and this time the objects are all the individual parts constituting the BOM as illustrated in Fig. 4.7.

4.3.3 Task Analysis

We are now able to analyse the tasks to establish their resource requirements as shown in Fig. 4.8. When all the tasks have been analysed in this manner it becomes possible to obtain the information on time and cost as shown in Fig. 4.8. The total sequence time is of particular interest as it enables the determination of number of workstations required to produce the product in volume.

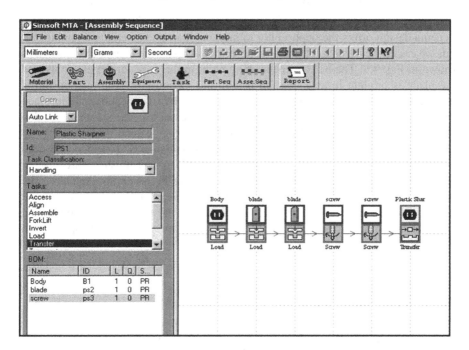

Fig. 4.7 The assembly task sequence

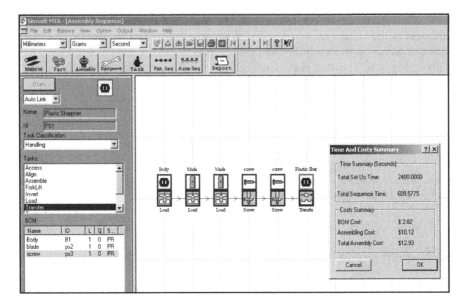

Fig. 4.8 The time and cost summary

4.4 Integrated Approach to Part and Assembly Planning

Schema design offers the opportunity to integrate part and assembly planning; for this we create the part manufacturing sequence alongside the assembly sequence as shown in Fig. 4.9. The body is the starting point in this case for the assembly and the blade is brought in as a finished part in the second task. The relationship is maintained by the connecting link that ensures a live connection of the blade manufacture. This brings in the time and cost values of it in the summary results. This demonstrates the capability of this methodology to extend process design to higher levels in order to allow 'what if' scenario assessments. The final cost does not change in this case whether the part is brought in as a finished part through BOM, or its manufacture is directly incorporated into the assembly sequence. The advantage of doing this is that various scenarios can be examined, for example the nature of tasks can be changed, different manufacturing equipment can be looked at, or the cost of raw materials or labour can be changed. It is a useful capability in this regard.

4.5 Process Knowledge

The process of creating data for parts manufacture and assembly provides us with linked files and these files can be copied and used for new products planning.

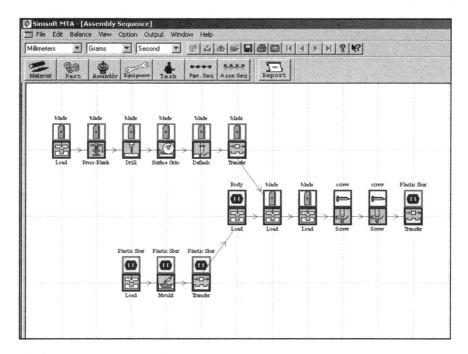

Fig. 4.9 An integrated perspective of part and assembly planning

The planner develops inference through experience which allows him to make rapid decisions on the best possible solutions. To structure this inference capability is not easy and the current state of art does not even allow us to attempt it, a little insight will help here. The distinction between different types of knowledge is that it can either be tacit, which means it is inherent to the people or the ability they possess, or it can be explicit, which means it can be structured or verbalised and therefore communicated to others. The tacit knowledge is inherently non-structured and cannot be recorded or easily transmitted to others and different capabilities of people reflect this, it is particularly noticeable in those who are regarded as gifted. We cannot develop such capabilities through education or training alone even if we try, such is the nature of it, organizations pay for it by salary means to those regarded as specialists. The explicit knowledge on the other hand offers different possibilities and is of particular interest to us here. The process narrative is an example of explicit knowledge and words and their associations give it a structure. We can go further and use visual means to convey its essence. At more advanced levels it can be represented as rules and algorithms and design for manufacture and assembly rules are examples of it. Explicit knowledge can also be represented by ontology; dictionary defines ontology as a set of entities presupposed by a theory or considered as specifications of concepts relating to a specific domain. The concepts outlined in this monograph centre on explicit knowledge, the narrative parsing and data relationships provide the structure and the interface

Fig. 4.10 The copy
capability at various levels

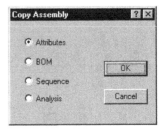

gives it visibility. This creates number of opportunities for process knowledge
management and we look at them in detail.

The unified and linked nature of the data files enables them to be copied and
modified for alternative assessments, this helps to reduce the planning times and
the costs of new process designs, and also allow 'what if' scenarios to be done
rapidly. This is often not possible through experience or through the use of
spreadsheet based systems. If we make a copy of the data file of plastic pencil
sharpener at assembly level it captures all aspects of it, including the manufacture
of parts. We can then explore the effect on time and cost of changing the body
material, for example to aluminium with similar overall dimensions. To do this we
create a part with aluminium as material and then create its manufacturing
sequence, we then substitute this part in the assembly BOM for the plastic
sharpener. Such a change requires live relationship among the variables involved
and the schema design allows this to be done at number of levels as illustrated in
Fig. 4.10.

4.6 Inferencing of Process Knowledge

The schema captures the association of all the tasks in a given manufacturing
activity and this association can be leveraged for inference purposes. For example,
the plastic injection moulding process has certain tasks and equipment requirements because of the nature of material involved. Similarly, the manufacture of
aluminium body has certain equipment and process requirements to complete the
part. Such associations are embedded in the narrative of process and the key words
can be used for inference. The first few action words always reflect the main
activity involved and can be used for search of a similar process in the database, as
for example in the use of Internet. Such inference enables the leveraging of what is
known and this saves time and cost. Over the years vast amount of knowledge is
developed and it remains in the memory of personnel involved, this is due to the
lack of suitable tools to capture it. The activity of part manufacture and assembly
planning is the interplay of materials, parts, assemblies, equipment and task
attributes and the amount of data involved is therefore large, hence the difficulties
of optimisations. The data is the final level of detail and enables the transformation
of process description into implementable concept. It can be extensive at micro

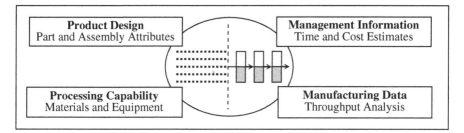

Fig. 4.11 Networking of process knowledge

levels, as for example in the development of tool paths for machining. It is also in the development of this data that the quantitative aspects of process design emerge and this enables the establishment of important parameters such as time and cost.

4.7 Networking of Process Knowledge

The practice of the old was individually centred and the practitioner recorded and used the same information, hence there was no loss of process knowledge. He understood the intent involved, even though the recorded information may not have been detailed enough. Such a practice is no longer possible because the recorded information has taken on a new meaning. The management tools now separate the application from files and this allows the application to be located on a server, this enables the information to be shared among many. This methodology provides for this as illustrated in Fig. 4.11.

Chapter 5
Tutorial on Part Planning

5.1 Introduction

The methodology outlined in this monograph is used in the teaching of undergraduate and postgraduate students and the feedback received has been encouraging. Students get to see the implication of their design decisions on cost in a more direct manner. This chapter and the next outline the tutorials developed through teaching. The software and to how to install it can be downloaded from the website http://www.simsoftks.com. There are a number of steps involved in part planning and they are outlined in Fig. 5.1. To experience them in detail we go through a simple exercise involving the manufacture of a pencil sharpener shown in Fig. 5.2.

Material Data	Part Attributes	Process Equipment	Task Sequence	Task Analysis	Process Plan

Fig. 5.1 The steps involved in part planning

Fig. 5.2 A plastic bodies pencil sharpener

S. Grewal, *Manufacturing Process Design and Costing*,
DOI: 10.1007/978-0-85729-091-5_5, © Springer-Verlag London Limited 2011

5.2 Material Data

Select Material in the Main menu window then right click to select New.

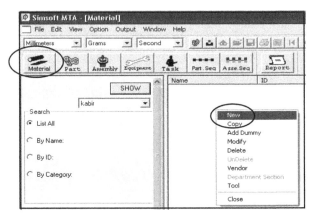

Fill and select details as per below, click OK.

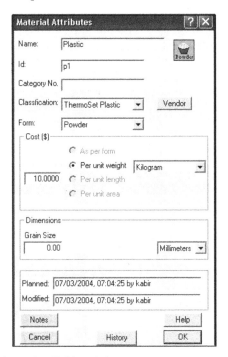

Close Material window via clicking 'x'.

5.3 Part Data

Select Part in the Main menu then right click to select New.

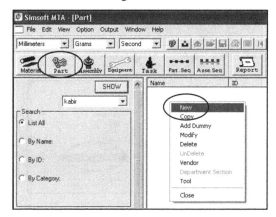

Fill and select details as per below, click OK.

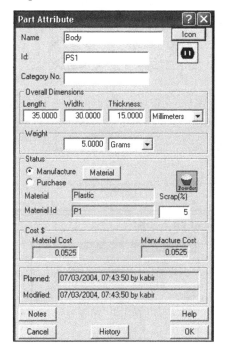

Close Part window via 'x'.

5.4 Equipment Data

Select Equipment in the Main menu then right click to select Department Section.

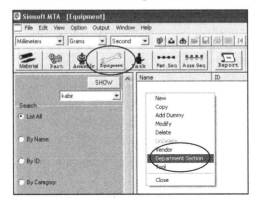

Click on New.

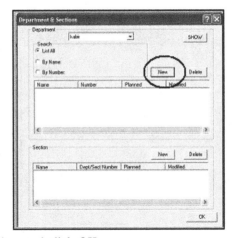

Enter details as below and click OK.

Select Plastic Department and click on New of Section.

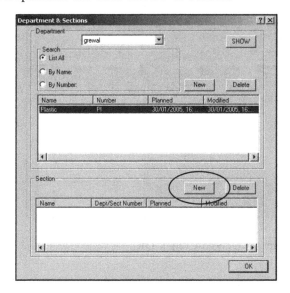

Enter details as below and click OK.

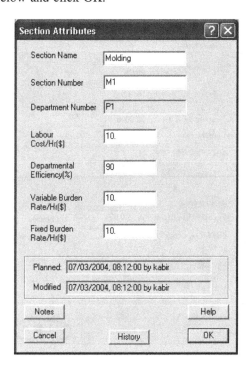

Close the Department/Section window.

Via Equipment button select New.

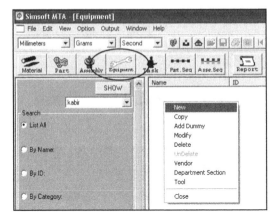

Enter and select details as below, click OK.

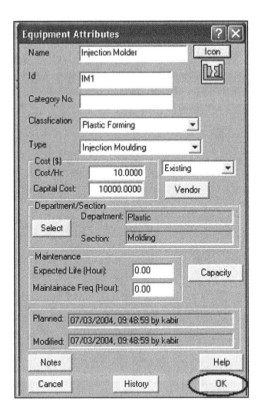

Leave the Equipment window open.

We now proceed to create information on the injection moulding tool.
Right click to open the sub menu again, select Tool.

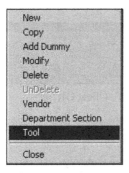

Click on New

Enter details as below. Select Icon. Leave Vendor details. Click OK.

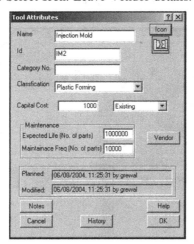

5.5 Task Sequence

Click on Part Sequence button.

Click on Open and select Body. Click OK.

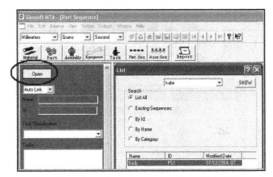

From the Task Classification drop down menu select Handling. In the Tasks menu select Load. Left Click at the desired location on canvas to place the task. Left click on the task and drag it to relocate if required. From the Task Classification drop down menu select Plastic Forming. In the Tasks menu select Mould. Click next to the load task to locate the new task. From the Task Classification drop down menu select Handling. In the Tasks menu select Transfer. Click next to the mould task to locate the new task. All tasks will have a green border at this stage, indicating they have not been analysed for their requirements at this stage. Save the task sequence by clicking the floppy disc symbol in the menu bar as shown below. Do this after the placement of each task as a safeguard; your effort will not be lost due to any computer system failure.

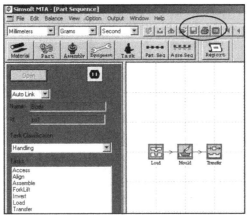

5.6 Task Analysis

Reopen Part Sequence window and via Open button select Body. Click OK. We now analyse each task. Double click on the Load task and fill in the details as per below. Select Manual. Click on Dept/Section button. In the Department window select p1. Under section select m1. Click on Cost Analysis button.

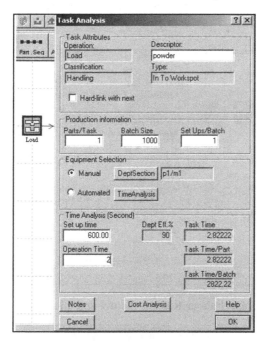

Fill in the labour rate as below. Click OK.

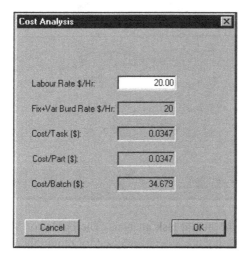

The border of the Load task will now change to black indicating the task has been analysed for its micro requirements. Click the floppy symbol in the menu bar to save the analysis. Do this after every task analysis. Double click on the Mould task. Fill in the details as below. Click on the Equipment button and select Injection Moulder. Click on the Tool button and select Injection Mould. Leave Check Capacity. Fill in the remaining details. Click on Cost Analysis button.

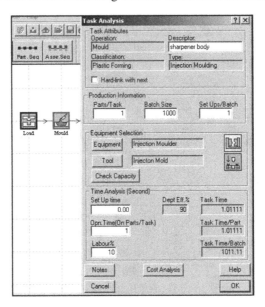

Fill in the Machine and Labour Rates as below and click OK.

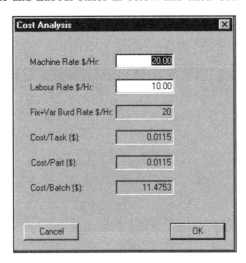

Click OK to accept the mould task analysis. The border of Mould task will now change to black indicating it has been analysed for requirements. Save the data via

the floppy symbol. Double click on the Transfer task and fill in the details as
below. Select Manual operation and under Dept/Selection select Plastic Department and select Assembly section. Click OK. Fill in the remaining details and
click on Cost Analysis.

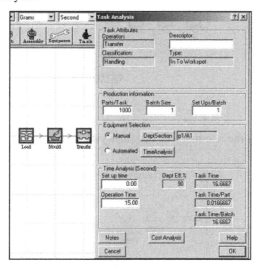

Fill in the Labour Rate as below and click OK.

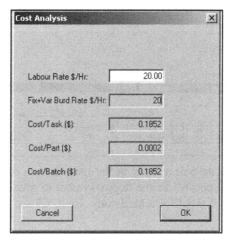

Click OK to accept the Transfer task analysis and save the data.
All tasks will have a black border now.

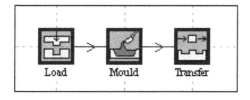

Right click on the canvas and select Time and Cost.

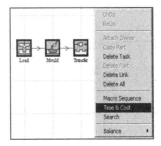

The window shows the summary of task analysis.

5.7 Results

Click on Reports button

Select Part Sequence tab. Select Body in the list and then select Time Analysis.
A detailed report will appear. Use the floppy symbol to save it in any format you
like or use the printer symbol for a hard copy. Click Close.

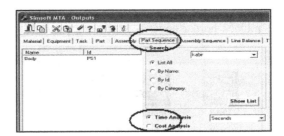

Chapter 6
Tutorial on Assembly Planning

6.1 Introduction

Assembly planning process involves a sequence of tasks that bring together all the parts. We look at this via the assembly of pencil sharpener. The steps involved are outlined in Fig. 6.1.

Fig. 6.1 The steps involved in assembly planning

6.2 Assembly Attributes

Select Assembly in the main menu window and right click to select New.

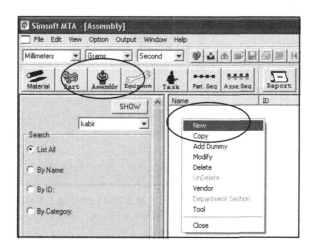

S. Grewal, *Manufacturing Process Design and Costing*,
DOI: 10.1007/978-0-85729-091-5_6, © Springer-Verlag London Limited 2011

Fill the details and select icon. Leave the Weight field at 0.00.
For Status select Assemble and click on BOM button.

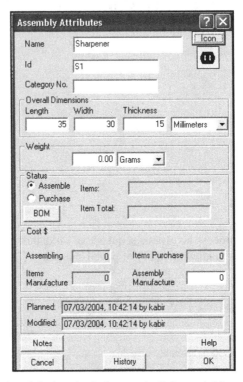

Select Body from the right hand window and click on Add.

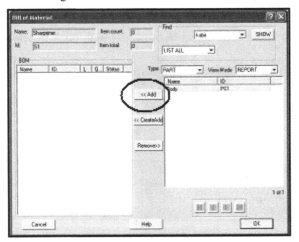

Select Add Weight & Cost. Leave Quantity at 1 and Level at 1. Click OK to accept
the data. The part will be added to the BOM window. Double click on it see the
attributes again.

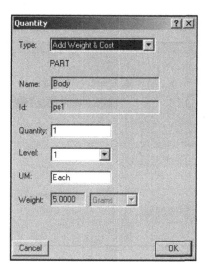

Select Simsoft as the user. Select Blade and click on Add. Change the quantity to 2. Leave level at 1 and click OK. Similarly select Screw and click Add. Change the quantity to 2 and leave level at 1. Click OK. The BOM is now complete and close the window. Click Update in the Assembly Attributes window.

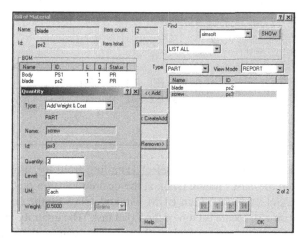

6.3 Task Sequence

Click on Asse Seq button.

Click on Open and select Sharpener, click Ok.

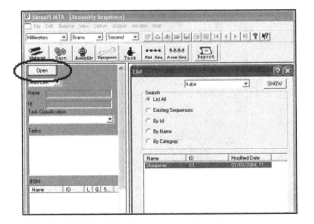

Sharpener icon will appear as shown below on the left.

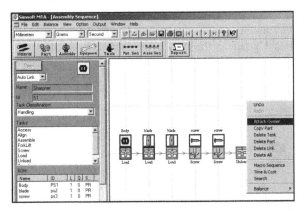

Under the Task Classification menu select Handling from the drop down menu. Select Load from the Tasks menu. Click at the desired location on canvas to place the 'Load' task. Select 'Body' from BOM window. Click on the Load task on the canvas. Body icon will locate on top of the Load task. The quantity value for Body under BOM will go down to 0. Select 'Load' task again and click next to the previous Load task on the canvas. A connecting arrow will appear. Select Blade from BOM and click on the new Load task. Select Load task again and locate next to the previous Load task. Select Blade from BOM and click on the new Load task. Select Screw task and locate it next to the previous Load task. Select Screw from BOM and locate it on top of Screw task. Select Screw task again and locate it next to the previous Screw task. Select Screw from BOM and locate it on top of the Screw task. Select Unload task and locate it next to the previous Screw task. Right click to open the options window and select Attach Owner. Click on Unload task and an image of the pencil sharpener will appear indicating the final assembly.

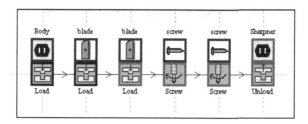

Save your data by clicking on the floppy symbol in the menu bar.

6.4 Task Analysis

Double Click on the first Load task. This will open the Task Analysis window as shown below. Fill in the details as shown. Through the DeptSection button select the Assembly department and the Sharpener section, it is under Simsoft user.

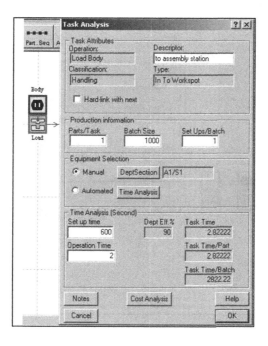

Click on Cost Analysis. Insert the Labour Rate and click OK. Click OK in the task analysis window also. The border of Load task will change to black indicating the task has been analysed. Save data via the floppy symbol.

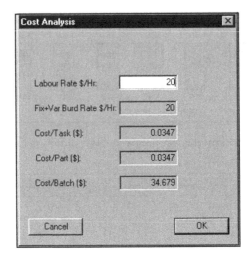

Double click on the next Load task. Fill in the details as below.

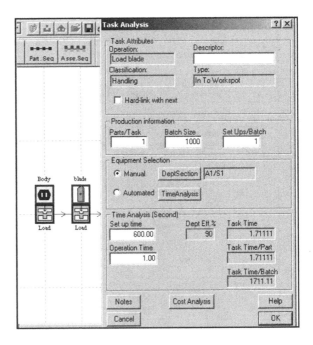

In Cost Analysis put 20 for the Labour Rate. Close all the windows. Save the data.

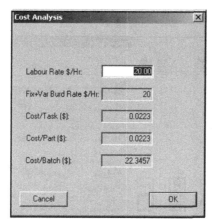

Double click on the next Load task and fill in the details as below. Note set up time is 0.

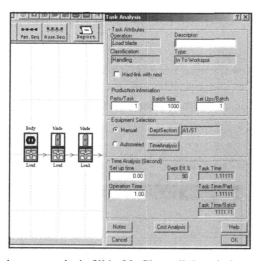

For labour rate under cost analysis fill in 20. Close all the windows and save the data.

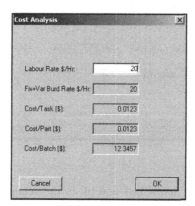

Double click on the next Screw task and fill in the details as below.

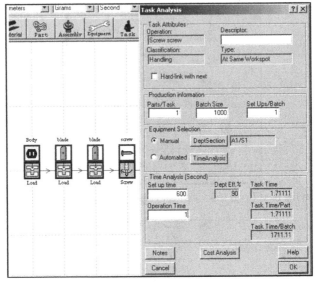

In cost analysis fill in 20 for the labour rate. Close all the windows and save the data.

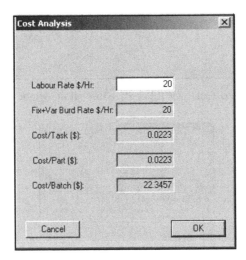

Double click on the next Screw task and fill in the details as below.

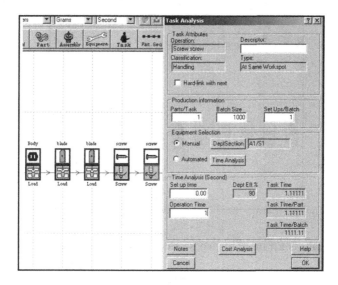

In cost analysis fill in 20 for the labour rate, close all the windows and save data.

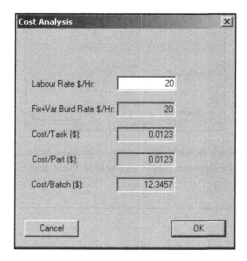

Double click on the Unload task and fill in the details as below.

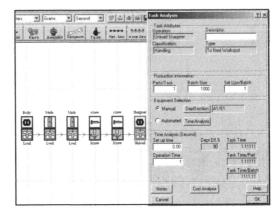

Under cost analysis fill in 20 for labour rate, close all the windows and save data.

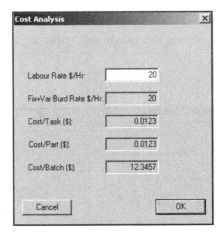

All tasks will have black border, indicating they have been analysed.
Right click to open the options menu and select Time and Cost.

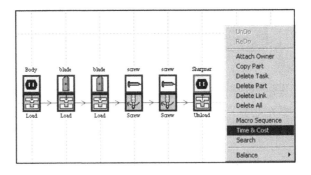

The Time and Cost Summary window shown below summarises the results of assembly analysis.

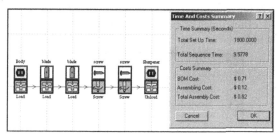

Assembly Attributes window should look like below. Click Update button if it does not. Note the Weight, Items and Item Total numbers, also the Cost information.

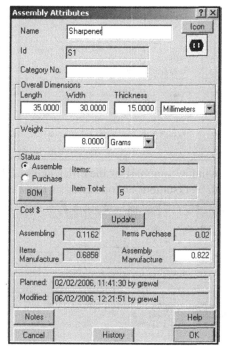

Close the Time and Costs Summary Window and Right click to open the menu. Select Balance followed by Sequence.

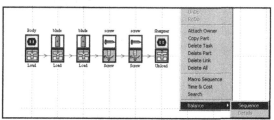

In the Line Balance Parameters window select Number of WS/Volume combination. Enter 1000 in the Volume section and 2 in the No of WS. Click on Show button. It will show the amount of time required to produce this volume at maximum workstation efficiencies. Click on Accept button. It will open the Line Balance Results window.

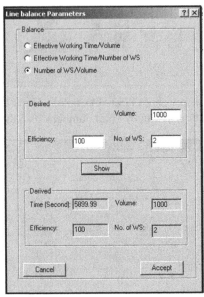

Line Balance Results window summarises the efficiencies of the workstations and their work content, both workstations are operating at 100% efficiency, click OK.

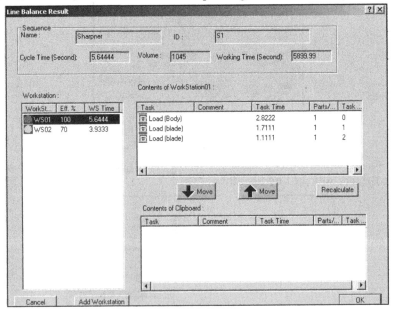

The graphical display indicates the combination of work tasks in the workstations to achieve maximum line balance. Once the number of stations and their action models are known it becomes possible to design the manufacturing system.

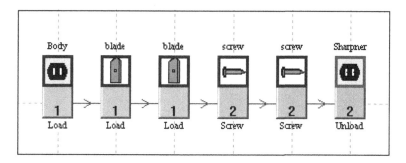

6.5 Results

Click on Reports button.

In the Outputs window select Assembly Sequence and output.
Time and Cost reports.

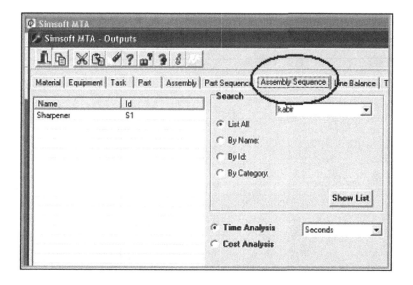

Chapter 7
Industrial Case Studies

7.1 Introduction

The research and development outlined in this monograph was carried out in close collaboration with industry and a wide range of products were tested, for feedback visit the Web site http://www.simsoftks.com. The case studies which follow demonstrate the industrial potential of this methodology; the first one deals with the assembly planning of an evaporative humidifier as shown in Fig. 7.1 and the second one deals with the costing of a moulding tool.

7.2 Evaporative Humidifier

The humidifier is composed of 60 parts and has overall dimensions of 273 mm × 259 mm × 207 mm and weighs about 3 kg. There are three stages involved in assembly planning:

Fig. 7.1 An evaporative humidifier

S. Grewal, *Manufacturing Process Design and Costing,*
DOI: 10.1007/978-0-85729-091-5_7, © Springer-Verlag London Limited 2011

Fig. 7.2 The handling
attributes data of waterfall
tank assembly

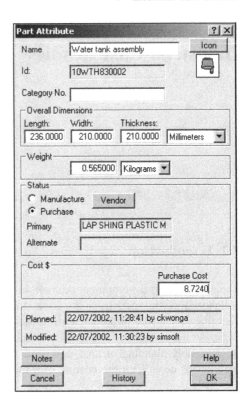

- creation of the handling attributes of the parts;
- the development of the assembly task sequence;
- the analyses of tasks to establish the resource requirements.

7.2.1 Handling Attributes

The handling attributes of the parts capture the broad parameters involved in terms
of their size and weight. Figure 7.2 illustrates the data for the water tank assembly.
The assembly attributes data of the humidifier is illustrated in Fig. 7.3. The cost of
items purchased comes from the bill of materials (BOM) and other costs can also
be noticed. The assembling cost refers to the assembly activity, the items manufacture refers to the cost of items manufactured in-house and the assembly manufacture costs refers to the total cost of manufacturing the product.

7.2.2 Assembly Sequence

Assembly is performed by sequence of tasks and this is determined by the
precedence requirement, which in turn is decided by the parts and the tasks to be

Fig. 7.3 The assembly
attributes data of the
humidifier

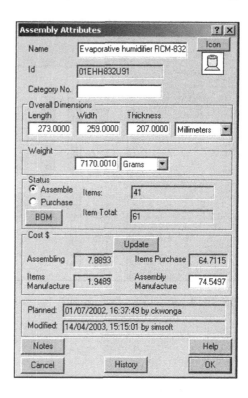

performed on them to enable the final assembly. The assembly task sequence
design capability of the software allows this to be done in a user-friendly
manner. You first select the task from the Task library and then the part on
which the task is performed. When all the parts in the BOM window are used up
in the task sequence, the generation of the sequence is then complete. For the
evaporative humidifier, the task sequence is long and a partial view is shown in
Fig. 7.4.

7.2.3 Task Analysis

Each task is analysed for operational requirement. The focus is on the tooling and
the time and this helps to determine the cost of the task. Figure 7.5 shows the
analysis of Load task and the data on labour and overhead cost. From these
variables the cost of carrying out the task is calculated. When all the tasks have
been analysed, it becomes possible to calculate the overall cost of assembly as
illustrated in Fig. 7.6.

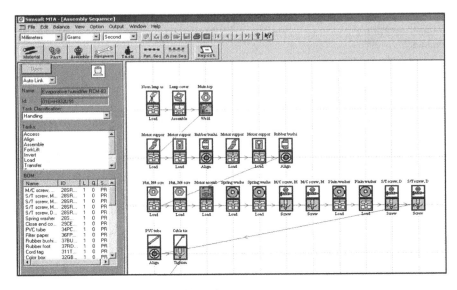

Fig. 7.4 A partial view of the assembly task sequence

Fig. 7.5 The analysis of Load task

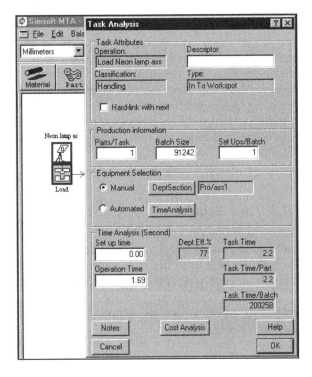

Fig. 7.6 The assembly time
and cost summary

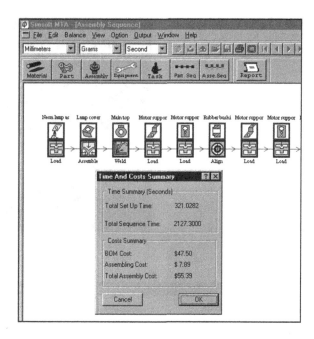

7.3 Costing of Injection Moulding Tool

We look at the costing of an injection moulding tool for the manufacture of
turbo fan grill as illustrated in Fig. 7.7. The front grill of the turbofan is a
plastic moulded part and the tooling for it is illustrated in Fig. 7.8. The

Fig. 7.7 Turbofan

Fig. 7.8 The moulding tool
assembly for the turbofan
grill manufacture

Fig. 7.9 Part and material attributes data for the cave insert

moulding tool is an assembly of 71 parts and the main cave insert part is
manufactured in-house, all other parts are purchased items. There are three
stages involved in the costing:

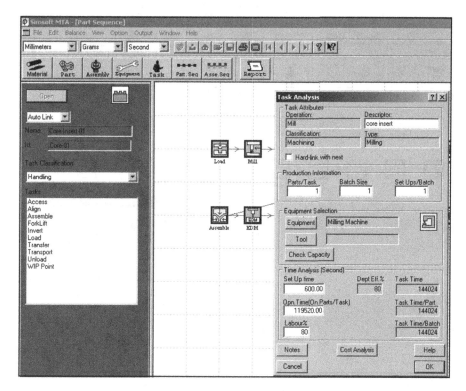

Fig. 7.10 The analysis of Mill task

- the creation of the materials and the handling attributes data for all the parts;
- the development of part manufacturing task sequence for the cave insert;
- the development of the assembly task sequence for the moulding tool.

7.3.1 Part Planning

For the moulding tool the most expensive part to be manufactured is the cave insert, which is a mirror image of the fan grill, as can be seen from Fig. 7.8. The cave insert involves complex manufacture and the material attributes and the handling attributes data of it are shown in Fig. 7.9. The manufacture of the cave insert part is performed by sequence of tasks and these tasks have precedence. The methodology of the software allows this to be done rapidly. The tasks are then analysed for their resource requirements and Fig. 7.10 shows the analysis of Mill task.

All the tasks were analysed in this manner and their times and costs aggregated as shown in Fig. 7.11. The cost of manufacturing the cave insert component is

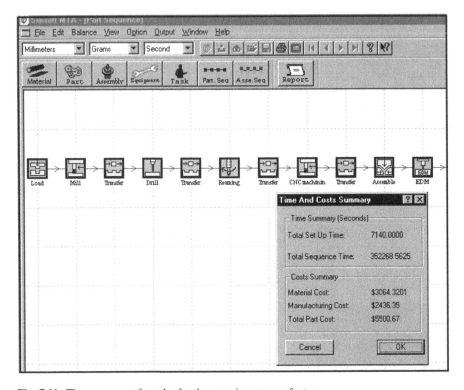

Fig. 7.11 The summary of results for the cave insert manufacture

carried into the BOM for the mould tool assembly. Any changes in the cost of manufacturing the cave insert component then reflect in the assembly cost as live relationship.

7.3.2 Tool Assembly

For the mould tool assembly, the handling attributes data are shown in Fig. 7.12. Such details were captured for all the parts, including the purchased ones. A section of the assembly task sequence for the mould tool is shown in Fig. 7.13, including the summary of cost.

7.4 Summary

This brings to conclusion all aspects of this methodology. It is hoped that it is of value to those interested in manufacture planning. The overall aim has been to

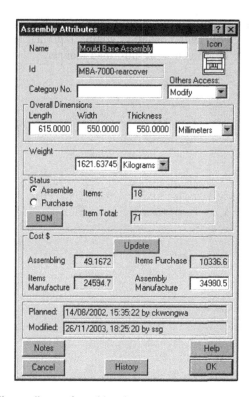

Fig. 7.12 The handling attributes of mould tool

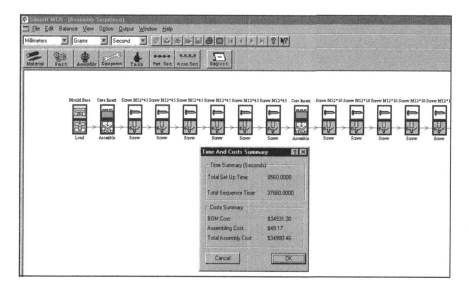

Fig. 7.13 The summary values for the tool assembly

advance the understanding and practice of manufacturing process design. The integration of this activity with costing is vitally important as it allows business decisions to be made in a much more informed manner.